About the Author

Garth Cameron is a lawyer, author, commercial pilot and flying instructor. He is also a glider pilot and instructor and has flown over 3,000 hours in many types of aeroplane and glider. He lives in Dunedin in the South Island of New Zealand.

HOW TO FLY A PLANE

You Can Experience the Freedom of Flight

Garth Cameron

RainShadow Press

RAINSHADOW PRESS
PO BOX 5441
MORAY PLACE
DUNEDIN 9058
NEW ZEALAND
garth@garthcameronlaw.co.nz
++ 64 27 622 8428

ISBN PAPERBACK 978-0-473-43263-8
ISBN EBOOK 978-0-473-43264-5

Contents

INTRODUCTION

Then it was intoxicating. The smooth takeoff, and the free feeling of having the world drop away. Soon after leaving the ground, they were crossing patches of stratus that lay in the valleys as heavy and white as glaciers. North for the first time. It was still an adventure, as exciting as love, as frightening.'

James Salter

My aim is to communicate to you the wonder and joy of flying and to do this by way of three threads to my story. Firstly, my flying experiences, secondly something of the technicalities and thirdly the best books by author-pilots.

Freedom is a basic human right and the aviation age has allowed humankind to access the ocean of air in a very liberating way. Most of the world's population has experienced the benefits of air travel but few of them have had the experience of piloting themselves. The freedom to move in three dimensions, and to see the world from above are experiences which have only been possible in the last few generations and I feel privileged to have had the opportunity to do so.

My total flying time as pilot-in-command is a fraction over 3,000 hours which is less than a lot and more than a little. Some pilots fly for years and accumulate less than 500 hours while career commercial pilots have between 10,000 and 20,000 hours in their logbooks, and some have over 30,000 hours.

You do not need a sophisticated aircraft to sample the unique experiences that flying offers and my flying has been in simple aeroplanes and gliders. It has been mostly in the daytime, clear of cloud and in sight of land or water. Nearly all of it has been in single-engine aeroplanes.

This book attempts to convey the challenges and pleasures of acquiring, developing, and using the skills required to fly small aircraft Each chapter describes a part of the process and the final chapter introduces the reader to some of the books I have enjoyed and explains why they are worth reading.

PROLOGUE. FIRST SOLO

Monday 10 November 1966, a few months after my 16th birthday. A grass airfield on a warm day with a light breeze from the north-east.

My instructor was an ex-Royal Air Force pilot who had defied the odds and survived 30 operations over Germany flying Avro Lancaster bombers in 1944-1945.

The place was a large airfield with three intersecting runways of mown grass. Once a busy air force and airline base, it is now home to an aero club, a gliding club and some private owners and a few topdressing planes.

The aeroplane I was flying was Cessna 150, a simple two-seat aeroplane of 100 horsepower. I had done all the preliminary training had done a dozen or so sessions of circuits, each one a take-off followed by a rectangular flight around the airfield and an approach and landing. To get the greatest number of circuits into the time available we did not stop but piled on the power after each touch-down and repeated the circuit. After the fourth landing, Bob told me to stop the aircraft but leave the engine running. He told me that I was to do my first solo and, after a few words of advice, climbed out, did up the seatbelt, closed the door and walked away without a backward glance.

I had thought about this moment for as long as I could remember, and any nervousness was obscured by the routine of taxing to the downwind end of the runway and doing my DVAs (drill of vital actions), a memorized checklist that covered the simple but essential items, a ritual that preceded every take-off.

The last item was to check that no aircraft was approaching, and I taxied onto the runway that faced east and lined the aircraft up.

Smoothly opening the throttle, the Cessna accelerated while I looked ahead and maintained a slight back pressure on the stick. When the aircraft was ready to fly she lifted smoothly off the ground and I let the aircraft accelerate to a safe climbing speed. At about 200 feet and 65 knots IAS (indicated airspeed) I retracted the flaps, reduced power and climbed out over the farms, farmhouses and roads of the Taieri Plain.

At 500 feet I turned through 90 degrees to the left and at then another 90 degrees in the same direction and levelled off at 1000 feet, flying parallel to the runway. Looking back and seeing the threshold of the runway to my left and slightly behind the wing I pulled the carburettor heat to hot and reduced the power while turning left until the runway was 90 degrees to my left. Two notches of flap and adjusting the trim wheel settled the airspeed at 65 knots.

A final turn to the left brought the runway threshold to my front and I pulled on full flap and let the airspeed fall off to 55 knots. As the airfield fence passed underneath, I reduced power and flattened out, holding the nose up until the rumble of the wheels told me I was back on terra firma.

CHAPTER ONE. BLAMING BIGGLES

Sunday 21 December 1969 was a warm day at the airfield with the grass runways, and there is a grainy colour photo taken on that day of me sitting in an open-cockpit biplane.

Captain W. E. Johns was the creator of a character called Biggles, an English pilot. Johns was a pilot in the Royal Flying Corps and Royal Air Force during the First World War and the character he invented, became an inspiration to several generations of boys, including mine. Johns knew what he was talking about because of his experience including flying de Havilland DH 4 bombers, surviving being shot down and spending time behind barbed wire as a prisoner of war.

The Biggles books started coming out in the 1930s and by the time Johns died in 1968, he had written over 160 books, including almost 100 Biggles books. Biggles was a young pilot in 1918 and did not seem to age much as the novels came out decade after decade. He flew in both world wars, small wars in fictional countries and explored jungles and deserts. A dastardly Hun named Erich von Stahlhein confronted him in book after book, until Biggles, in *Biggles Buries a Hatchet*, rescued him from a Soviet labour camp on Sakhalin Island after World War Two. Not all the plots were credible and from the 1960s onwards some people characterized the Biggles books as jingoist, racist and misogynist, criticisms that have the ring of truth about them. All in all, they were ripping yarns and I read over 50 of them.

V. M. Yeates auto-biographical novel *Winged Victory* was published in the early 1930s and Yeates characterized the first Biggles book, *The Camels are Coming*, as 'superbunk'. Yeates' objection was probably that Johns failed to describe the heavy drinking and the traumatic and disabling fear that many pilots felt. His novel provides descriptions of air combat and ground strafing by pilots flying Sopwith Camels in 1918 that have never been surpassed. Ernest K. Gann's *In the Company of Eagles* is a tale about two fighter pilots in 1917, one French and one German. The plot is credible, and the flying and air fighting are first class. Gann was not a war veteran, but he was a 10,000+ hour airline pilot and had flown open cockpit biplanes in the 1930s. Both these books are about people as much as about combat flying and have believable plots and excellent characterisations.

The closest I got to flying like Biggles was two years after getting my Private Pilot's Licence when I did a type rating on a de Havilland DH 82a Tiger Moth, the aeroplane I was sitting in, in the photo. The Tiger is a two-seat open-cockpit biplane with a tail-skid instead of a tail-wheel. More than 300 of them had been manufactured in New Zealand and the type had served throughout World War Two as the Royal New Zealand Air Force's standard basic trainer and many went to aero clubs and topdressing companies afterwards. The Southern Districts Aero Club operated it, an old aeroplane and not yet a vintage type. I see from my logbook that I did my first solo on the type on 20 December 1969, at Charlton airfield, just outside Gore.

It was summer, but I still needed warm overalls and a leather helmet and goggles. The pilot (or pupil pilot) sat in the back seat and had a poor view ahead, and the pilot had to swing the nose left and right to clear the area ahead while taxing and to look out to one side during the take-off. The aircraft would spin off a stall and keep spinning until the correct control inputs were applied; throttle closed, full opposite rudder, stick forward until spin stopped, level the wings, apply power and climb away. Before I soloed, I had to demonstrate adequate recovery skills and had to do spins solo. I remember a film of perspiration inside my helmet. One day I decided to see how high the aircraft would climb and got up to 8,500 feet before giving up, if I had been patient it would have gone well over 10,000 feet.

World War One flying books planted a seed that my father nurtured. He had served in the Royal New Zealand Air Force from 1940 to 1946. He trained as a fitter 2E, an expert in aero engines, a skill that took several years to acquire. He managed to avoid doing anything heroic and his service in Fiji entitled him to membership in the RSA (Returned Services Association), an important thing, as men of his generation were divided into those who had served overseas and those who had not. RSA membership was a subtle but important plus for veterans when they returned to civilian life.

My father did not fly for more than ten years after the War as he had seen so many crashes he judged flying to be a hazardous activity. The RNZAF Curtiss P40s are a striking illustration of the high accident rate. New Zealand received just under 300 of them and 76 were lost in accidents in New Zealand another 76 in accidents overseas. By way of contrast, only 20 were lost in combat. Who would say that my father's assessment was wrong?

Peacetime aviation was far safer than wartime and around 1961 my father chartered a Cessna 172 to take him to inspect some farm machinery. The return flight took one and a half hours while the drive would have been ten hours, before long he was learning to fly.

CHAPTER TWO. FIRST FLIGHT

I had my first flight in January 1961, in the world's most popular aeroplane, a Cessna 1972,

The CFI and my father sat in the front seats while I sat in the back. The CFI started the engine and taxied towards the active runway and I noticed that as the aircraft rolled over the slightly uneven grass surface of the taxiway there a squeaking sound from the joint between the plexiglass windscreen and the sides of the cabin, a sound that was slightly disconcerting but normal. On the ground the rudder pedals steer the nosewheel and pressing the tops of them produces differential braking. The 172 stopped just off the end of the runway, with a view of the approach and take-off areas while the pilot ran through his pre-takeoff checklist. I could see him running his eyes and fingers over the gauges, switches and toggles. The flaps were extended to 10 degrees and the ailerons, elevators and rudder were moved through their full travel. The aircraft taxied onto the runway, ran forward and the pilot applied full power, seconds later it was airborne and climbing away. This is an experience I have had more than 15,000 times since and I am still awed by it, the magic of flight.

Cars are produced in their millions, but an aircraft is successful if it sells in its thousands. The Cessna 172 was first produced in 1956, more than 44,000 have been sold and it remains in production today.

Before each flight, the pilot walks around the aircraft doing the pre-preflight and checking what must be checked. The first step is to check that the ignition and parking brake are off, the master switch is flicked on, the flaps lowered, and the master switch turned off. The pilot takes the fuel dipstick out of the pocket behind the pilot's seat. Fuel quantities are checked visually by dipping the stick into both fuel tanks and the oil by a built-in dipstick. I always note the fuel and fuel endurance on the back of my left hand. The pilot circles the aircraft in an anti-clockwise direction checking more than 40 items, including the pitot head, static vent, control surface travel, their hinges and mass balances. The words are incomprehensible to the beginner but have become second nature by the time the beginner has become an experienced pilot.

The cockpit has two adjustable seats in the front and a bench seat for two people in the back, aft of the rear seat there is a luggage compartment. There is a set of rudder pedals and an oblong control yoke in front of each of the front seats. There is a row of controls along the bottom of the instrument panel, including a primer, ignition switch, throttle, mixture control, carburettor heat control and fuel cock. The pilot or pupil sits on the left and the instructor on the right. The instructor can reach all the controls but must lean over to the left to use the ones in front of the pupil.

The instrument panel contains instruments monitoring the health of the engine and the speed, altitude, heading and attitude of the aircraft. The airspeed indicator is colour-coded. The white arc is the speed range at which it is safe to fly with the flaps extended while the green arc marks the speed range with which it is safe to use full deflection of the controls or to fly in turbulence and the red line is the never exceed airspeed.

All pilots learn that the senses are unreliable when the horizon cannot be seen, in cloud, fog or white-out. Without blind flying instruments and the training to use them a pilot will become disoriented. A pilot may do a climbing turn, losing airspeed, stalling and falling off into a spin, or a descending spiral with increasing airspeed. Either way, the lesson is, stay out of cloud. Like all pilots, I did some simulated instrument flying, not so I could fly blind but to drive home the lesson and help me to stay orientated when the horizon was indistinct.

In addition to the airspeed indicator, there is an altimeter, artificial horizon, turn and slip indicator, gyro direction indicator and compass. It counter-intuitive that, on a clear day, outside controlled airspace and away from airfields an experienced pilot could fly a 172 with all the flight instruments blanked out.

Creating and flying an aeroplane requires a unique synthesis of imagination and creativity, aerodynamic and engineering knowledge, hand-eye skills, judgement, and self-discipline. A flight is an adventure and is an event with the potential to satisfy a man-kinds craving for the freedom to go anywhere, anytime, a little danger, the acquisition of meaningful skills and aesthetic fulfilment.

A simple aeroplane like the Cessna 172 is a vehicle which can train pilots and go almost anywhere almost any time. It is fitted with dual controls and can be used for the dual and solo flying required from first flight through to first solo, private licence, commercial licence, instructor rating and instrument rating. It can carry the pilot and three passengers, with room for a little baggage. With a cruising speed of around 100 knots and an endurance of four hours, it can (allowing for a half hour reserve) cover 350 nautical miles at a time if there is no headwind. With two refuelling stops, it can easily cover over 1,000 nautical miles in a day. The 172 can carry a doctor and a patient on a stretcher, an aerial survey camera, a load of skydivers or a mass of courier parcels or tow a glider behind it.

The striking thing about the 172 is how simple it is and how little it changed from 1956 to 2018. The primary structure is less complicated than a modern car and the engine is like that in a Volkswagen Beetle, albeit more powerful and reliable. Its aerodynamics are simple but provide easy handling and efficiency, with only 160 horsepower to play with. It is notable that a sky full of 172s in 2018 is like having roads full of Morris Minors, 1956 Fords, Vauxhall Velox's or Volkswagen Beetles.

General aviation has an image problem and the media gives undue prominence to accidents to small aircraft. A head-on collision on the road claiming several lives is just a news item but reports of a fatal accident to a small aeroplane are often sensationalist.

I am often asked 'is it safe' and I explain that it depends on the pilot, the aircraft and the context. An experienced pilot doing circuits in fine weather is safer than driving to the airport. Flying in well maintained, factory-built aircraft such as Cessnas and Pipers is relatively safe while flying in basic microlights which may have shoddy maintenance can be risky. A low time pilot flying in poor weather, in mountainous terrain is an accident waiting to happen but a high-time pilot flying in good weather is probably as safe as any other day to day activity.

Good pilots make safety a prime consideration. They always check the weather, plan the flight on paper, if flying away from the airfield, or in their heads, if it is a local flight. They do the pre-flight described above and go through checklists before taking-off, before doing air-exercises, such as maximum rate turns, stalling or aerobatics. When near an airfield they are very air-space conscious and keep an especially good lookout. On the downwind leg of the circuit, they check items including brakes and fuel quantities. The same pilot's study accident reports which are circulated to inform and educate all pilots. They analyze each-others flying and always try to learn more, fly better and be safer.

Another question I am asked is 'should I fly with an inexperienced pilot'. If your pilot is hiring his or her aircraft from a flying school or aero club and the aircraft is a Cessna or Piper or other factory-built aircraft and each flight is authorised by a supervising instructor, you can fly with a reasonable degree of confidence. Don't be shy about asking your pilot about his or her experience.

CHAPTER THREE. LEARNING

In late 1965 I started to learn to fly in a Cessna 150 a machine slightly smaller than a 172, with two seats and a 100-horsepower engine. There is a standard teaching and learning sequence in initial flying training, starting with the effect of controls and ending with take-offs, circuits and landings. In between these lessons you learn to fly straight and level, climb and descend, do gentle turns and stalls and recovery from stalls.

I was not eligible to fly solo until my 16th birthday so the training flights were widely spaced and interspersed with passenger flights with other pilots, including my father.

Dad had become hooked after the charter flight to inspect the machinery and had started to learn to fly. After the first few lessons and passing the written exams he took two months off work and flew every day until he had passed his private pilot's licence flight test. The first I knew of this was when he arrived home with his licence. We often flew together, first in hired aircraft, then in the family Piper Tri-pacer.

Cross-country flights are part of every pilot's training and require skills not needed to drive from place to place, following road-signs and roads. Student pilots find that the ground often looks different from above and that it is all too easy to get lost. My early cross-country flights were made decades before GPS made navigation easy and, in a time, when beginners did not use radio navigational aids. They were additional proof that air safety required a sound plan and attention to detail. The planning started with drawing a track line on a chart and measuring the direction with a protractor and the distance with a ruler calibrated in nautical miles. The cruising speed of the aircraft was known as was its endurance. The weather forecast provided wind direction and speed. All of this was transferred to a flight log sheet. The wind and the variation (the difference between true north and magnetic north) was applied to an E6B computer (a circular slide rule) to turn the raw information into a heading and time for each leg of the flight.

Cross-country flying with a map, compass and watch was infinitely more demanding than flying by a GPS but most pupils and pilots, who are sometimes uncertain of position, end up where they intend to go. Even today, pilots must know how to navigate with map, compass and watch, as insurance against the day that their GPS fails.

One of the aircraft my dad flew was a de Havilland Chipmunk, a type used by air forces as a trainer and operated by the aeroclub as an aerobatic aeroplane. It looked a little like a World War Two fighter with a tailwheel undercarriage, a sliding canopy covering two seats in tandem and a lean, mean look. The back seat was a little drafty and I was unimpressed at the time. Now-days they are often called a poor man's Spitfire, combining great strength with light and well-harmonized controls, ideal for aerobatics and generally throwing an aircraft around. I wish I had one now.

My first exposure to aerobatics was in a Chipmunk, owned and flown by Jack Hanlon, a successful structural engineer and the ultimate aviation enthusiast. In an era when few people owned an aircraft, Jack owned the Chipmunk, a Tiger Moth, a Slingsby Eagle two-seat glider and a Slingsby Dart single-seat glider. In addition, he owned and flew New Zealand's first modern hot air balloon. I had done about 50 hours when he invited me to sit in the backseat while he did some aerobatics. Five minutes' worth would have been fun but 25 minutes of loops, rolls, stall turns, and the like had me on the verge of vomiting. I have never thrown up in an aircraft, but that flight brought me as close as I have ever been. It also taught me a lesson, never throw an aeroplane around with an inexperienced passenger in it.

I passed my private pilot's licence in late 1967 when I was still in high school and my father was my first passenger. Dad had encouraged me to fly and had paid for my first 100 hours but would not let me ride a bicycle or motorbike. In retrospect, I think there was less chance of a prang with two wings than two wheels.

I often took my licence out, just to look at it, a document printed on cartridge paper, with a mauve strip across the top left-hand corner: New Zealand PRIVATE PILOT'S LICENCE No. 6448 dated 11.10.67. As icing on the cake, it came in a passport size green folder with gold lettering on it. I carried it everywhere and took to sneaking a look at it during boring classes.

My first job paid $22 a week, which was not much, even in 1969. Flying was fun, but it was not yet completely satisfying, that would only come with experience and I looked for ways to build hours and skills without paying for them.

CHAPTER FOUR. LOGBOOKS AND TAILWHEELS

I remember the day in 1965 when my father gave me my first logbook, saying that I would never fill it, not knowing that I would, and another three like it.

A flying logbook becomes a pilot's treasured possession, the proof of his or her experience and more than a mere series of brief notes in clipped formal language. A line in a logbook can be evocative and recall a complex memory. Mine was light blue with dark blue spine and corners, a version of the RNZAF logbooks that many of my flying instructors had filled during World War Two.

When I left school, I had 85 hours and 50 minutes flying time, a little over half of it solo, this was not very much. I needed many more hours if I was to qualify for a commercial pilot licence and flying instructor rating. Sometimes I would persuade friends to come flying with and share the cost and I slowly built up my hours. I was doing some gliding and in 1969 I added a towing rating to the green folder and started to build flying hours and skills by flying the gliding club's Piper Super Cub tow-plane.

Not only were the tow flights short, meaning many take-offs and landings were packed into each hour, but the tow plane was also a tail-wheel type. Most aircraft have tricycle undercarriages with two main wheels and a nose wheel while the Cub had two main wheels and a small tailwheel. An aircraft with a tail wheel or tail skid had some advantages and significant disadvantages. It is more suited to rough and short airstrips while it is more difficult to taxi and take-off and land, particularly in a crosswind. If directional control is lost on the ground, a ground-loop is the result, with the aircraft chasing its tail. In a ground-loop, the aircraft rolls in the opposite direction and the wing and undercarriage can suffer expensive damage. If the ground-loop is violent the aircraft can dig a wing into the ground and capsize onto its back, damaging both the aircraft and the pilot's self-esteem.

Tailwheel aircraft required more skill to fly and flying the Cub was a pleasant challenge. In due course, I would fly open cockpit aeroplanes with tail skids in place a wheel and no wheel brakes. The skid let the aircraft weathercock into the wind and, with no wheel brakes to correct the yaw the only way of keeping straight was to increase the power. Added power increased rudder authority by blowing more air over it but also increased speed and the aircraft would bowl along with the pilot looking for a safe place to reduce power and turn into the wind. When aero clubs used Tiger Moths utilization rates were low because the wind was often too strong to allow inexperienced pilots fly and there was a high dropout rate of pupils who could not acquire the skills to fly them. When aero clubs replaced the old aeroplanes with modern tricycle undercarriages utilization rates shot up, more pupils stayed on and flying became cheaper.

I made myself available anytime and flew the tow-plane often, doing anything between 5 and 15 flights a day, 28 on one memorable afternoon. Soarable ridge lift and wave lift require high winds and meant that many of my flights were in strong winds, requiring a certain finesse on take-off, landing and particularly when taxiing across the wind. In very strong winds I would land and wait until club members arrived to hold on the wing struts and tail and prevent the Super Cub from blowing over while it was walked to the hangar.

Hours can be a poor measure of pilot experience and skill. The hours I spent sitting in the right-hand seat with my pupils doing circuits, were low-quality hours. The glider towing was of high quality. Flying a tricycle undercarriage aircraft with take-off and landings into a light and steady winds do not teach you much but landings and take-offs with a strong, gusty crosswind are both demanding and valuable experience. Hours spent machine-minding in cruising flight are just hours but the same time flying a fully loaded Cessna 185 into and out of a short and rough airstrip is much more meaningful.

As well as building hours I studied for the written exams for a commercial pilot licence. The exams were in; meteorology, technical knowledge, principles of flight, aviation law and navigation. None of it was rocket science but there was a mass of facts to absorb and apply to day-to-day flying.

One day in June 1975 I received another buff envelope marked HMSO and inside was a document, with a blue strip diagonally across the top left part of the page. It read: 'Civil Aviation Division, Ministry of Transport, New Zealand, COMMERCIAL PILOT LICENCE, Nr. 2976, 3 June 1975. There are many such licences, but I had aspired to hold one since I was a boy, so it was a special day for me.

I don't fly much these days, but when I do, I try to pack in as much as possible into the time. I could cruise around and only do one take-off and landing, but I don't. I write out my checklists before getting into the aeroplane and do a careful walk-around. You can do a lot into 45 minutes; steep and maximum rate turns, stalling clean and with flap and power on and several circuits including a flapless landing and a short take-off and a short landing. This is fun as well as meaningful.

CHAPTER FIVE. LANGUAGE

The international language of language is English, but aviation also has a language all its own using unfamiliar words, abbreviations, and colour codes. A Cessna 172 has an instrument panel that is simple compared with advanced aeroplanes, but a beginner sitting in the cockpit for the first time has a lot to learn about how to read and interpret them.

Aviation uses a curious mix of metric and imperial figures. Speeds are expressed in knots, distances in nautical miles and altitudes in feet. Runway dimensions and visibility distances are measured in meters. Oil quantities are in US quarts and petrol quantities may be measured in liters, imperial gallons, or US gallons. Fuel contents gauges are unreliable, and it is standard procedure to dip the tanks with a carefully calibrated dipstick. Occasionally you will come across an airspeed indicator calibrated in miles per hour. All pilots need to be alert to the danger of mixing units. If you assume that the gallons are imperial, but they are the smaller US gallons you risk sudden silence from the engine and embarrassment when you must walk home.

The abbreviations and numbers used in aviation include three-figure code groups such as QNH followed by a number (the setting of the altimeter so it indicates the aircraft's altitude above sea-level) and QFE followed by a number (the setting of the altimeter so it indicates the aircraft's altitude above the surface).

The abbreviations AD, AGL, AMSL, AQP, ATZ, BTPD, BTPS, CRM, DA, DH, EET, ELT, ELT(S), EPIRB, ETA, ETOPS, FAR, GAA, GPS, MDA, MEL, MNPS, MTOPP, NZPMA, NZTSO, PDZ, PLA, QFE, QNH, RNAV, RNP, RVSM, SFAR, STPD, TCAD, TCAS, TSO, V1, V2, VEF, VS and VSM were all inserted into the Civil Aviation Rules (New Zealand) at one time and represent the tip of a very big iceberg. Abbreviations which baffle the beginner soon become familiar and is normal to think of weather as being VMC or IMC and flight being in accordance with VFR or IFR requirements.

If you fly for long enough converting the abbreviations into words become second nature and it is rather fun learning them.

Flight instruction is weather dependent, but bad weather becomes a time to learn or brush up on the theoretical side of flying. I always encouraged pupils to spend time in the cockpit just getting used to the environment and identifying the instruments and switches and toggles and running through the checklists that every pilot must memorise. In the USAF it was normal for trainees to learn the checklists so well they could recite them in a loud voice while running and being distracted. Good advice for any pilot. In the RNZAF pupil-pilots must keep a meteorological logbook recording conditions every day and this helps them to become weather conscious.

Civilian flight training standards (at large flying schools and aeroclubs) have steadily improved over the last 30 years or so. When I was active in flight training the emphasis was on getting the pupils off solo on each stage of training, while now the emphasis is dual instruction with just enough solo to satisfy the licensing requirements. A pilot who trains full-time has an advantage over the pupil who must fit the training into work and family obligations.

CHAPTER SIX. THE AIR

The air appears to be insubstantial as you can't see it or smell it and it is only 1/800th the density of water but cleverly shaped wings hold aeroplanes in the air and knowledge of the basics of aerodynamics and meteorology is essential to safe piloting.

The temperature, pressure and density of air changes from place to place, time to time and with altitude. Designers and pilots refer to a standard atmosphere with well-defined temperature, pressure and density when designing and operating aircraft. The performance figures are adjusted to those which would exist if the conditions were identical to the standard atmosphere. If the air is cold and dense the take-off distance is reduced, and the rate-of-climb is improved. If the temperature is high and the density of the air is reduced the take-off distance is increased and the rate-of-climb is reduced.

Trainee pilots soon learn that aircraft take-off and land into the wind because this reduces take-off distance and improves the angle of climb. On landing into-wind, the angle of approach is steeper, and the groundspeed is reduced, resulting in a shorter landing distance. Taking off or landing down-wind reverses these effects and may result in disaster.

Gliders aren't silent, but they are quiet, and this is one of the many things that make gliding a special experience. They are also the ideal aircraft in which to enjoy and understand what goes on in the ocean of air. Designers strive to give their sailplanes the lowest rate of descent and the maximum glide angle. A Cessna 172 with the engine idling glides at an angle of about 12 to 1, that is 12 feet forward for every 1 foot down. The best modern sailplanes have glide angles of from 50 to 60 to 1. My first glider flights were in a machine with an 18 to 1 glide ratio.

I started gliding in 1968, and although I only did 169 flights and accumulated less than 60 hours, I learnt a lot about the air and the weather in the process.

I started gliding because sailplanes were so beautiful, but my first gliding lessons were in an angular machine which would not have been out of place on a gliding site in the mid-1930s. It was a Slingsby T 31 with two seats in tandem beneath a high wing braced with two struts on each side. It had recently been refurbished and a clumsy looking plexiglass canopy had been replaced by two small clear windshields, one in front of each cockpit. The open cockpits were not a problem, even in winter, because airspeeds were low and training flights were brief. Each flight consisted of an aero tow by the Club's Piper Cub tow plane to a height of between 1000 ft and 2000 ft and my logbook records 32 flights of between 8 and 15 minutes each before soloing for a glider for the first time.

I had also been inspired by Terence Horsley's book *Soaring Flight* is a delightful description of the sublime pleasures of soaring. Horsley's description of a flight in smooth ridge lift late in the day is entrancing. It is also a training manual and a description of gliding and soaring as it was on the outbreak of World War II. The book made me think about branching out from flying to gliding.

Simple gliders are straightforward to fly but there is only one chance to make an approach and landing. Part of the technique to arrive at the right height and position for a safe landing is to aim at a point about 1/3 of the way down the landing area and then steepen the glide by using the dive brakes. The T 31 was fitted with a primitive sort of dive brake called a spoiler. These spoilers were two devices each about 750 mm wide and 500 mm deep and were mounted in recesses on the top of each wing. They were held flush with the upper wing surface by springs. If the glider was too high the pilot opened the spoilers by pulling a knob and maintained speed by lowering the nose. This caused the glide to steepen without accelerating. If the glide became too steep the pilot eased the pressure on the spoiler knob, the springs closed them, and the glide flattened. The landing was a straight-forward affair of flattening out and holding enough back pressure on the stick, so the machine touched down on its single wheel in a level attitude. There was no wheel-brake, but the landing distance could be shortened by pushing the stick forward and putting the skid on the ground. The extra drag pulled the glider up short.

Training flights were all a tow to altitude followed by a descent to the downwind leg of the circuit (parallel to the landing strip) followed by the approach and landing. My first soaring flight happened one day when there was a westerly breeze at right angles to a line of low hills near the airfield. We arrived on the windward side to find that there was just enough lift to maintain height in. A glider soars when it encounters air rising at least at the rate that the glider is descending through the air. The glider is still descending relative to the air but is not descending relative to the ground. If the air is rising at more than the glider's rate of descent it will climb. On this day there was just enough lift to stay aloft for about 30 minutes and we happily tracked back and forwards along the windward side of the line of hills to stay in the lift until the wind dropped slightly and the rising air would no longer keep us aloft.

Glider pilots can soar by understanding the signs which mark areas of soarable air. All pilots know about the clouds; stratus, nimbus, cumulus, cumulonimbus, the high cirrus, which exists in a realm so high and so cold that they are made up of ice crystals. The almost myriad types of cloud indicate the types of weather on the way and are often beautiful to behold, so much so that there is a Cloud Appreciation Society in Great Britain. Some of the technical charts showing cloud types and characteristics are as beautiful as any painting.

Away from the coast thermals will form on hot days with light winds. Differential heating causes bubbles of heated air to break away from the ground and they rise fast enough for a sailplane to climb by circling in them. Often the thermals are marked by puffy clouds at their tops and pilots head for them to find the lift. On a cross-country flight in thermals, pilots will gain height in a thermal and then glide to the next, circle, climb, and repeat the sequence until they have reached their destination. Thermals are most prevalent, strongest, and most reliable in places like Texas where extraordinarily long flights have been made since the 1950s. I soared in thermals from time to time and on one occasion I thermaled dual in a two-seat Slingsby T 53, an all-metal machine that creaked in turbulence, a noise that did not inspire confidence.

About 10 nautical miles south-west of the airfield were a range of hills 6 miles long and about 3000 ft high. When the conditions were right (strong south-west wind and a high lapse rate), a standing wave formed downwind of the range. Imagine water flowing over a line of rocks across a stream. The water goes over the rocks, down and then smoothly up. Magnify that effect 1000 times and you have what happens in the air. This is called primary, secondary, and tertiary wave. At the top of each curve is a smooth lenticular cloud. The lens-shaped cloud is stationary despite the wind as it is constantly forming at its upwind edge and dispersing at the downwind edge. In front of each of the waves is a band of smooth air rising at many times the rate of descent of my glider.

The wood and fabric Slingsby Skylark IV I am about to fly has been superseded for contest flying by fiberglass sailplanes, but I love it. The fuselage, fin, and rudder are a mellow red and the wings, tailplane, and elevator are cream. The cockpit provides a semi-reclining seat and the 18-meter wings are at eye level with the leading edge just behind my head. This and the flawless Perspex canopy provide a breathtaking view and the view is another reason for going gliding. The structure is wood with a smooth plywood surface and only the nose is fiberglass.

The wind on the ground was a gusty 25 to 30 knots and the airspeed indicator registering 25 to 30 knots although I am stationary. It is strong enough for me to keep the glider's wings level with the ailerons even though I am not moving.

I run through the short but essential pre-takeoff checklist as the tow plane lines up in front and am offered the rings on the end of the tow line. We check the release, hook on and my friend goes to the port wing tip. I say, 'take up slack' and he waves his arm back and forwards underarm, so the tow-pilot can see him in his mirror and roll forward until the line is taut. 'All out' causes him to wave his arm back and forward over his head and I hear the Cub go to full power and I start to move. In this wind, I am airborne in 50 meters and hold a position in the high tow position out of the Cub's slipstream. Soon we are climbing to the south-west in rough air heading towards the area where the wave awaits. I know what I am in for and have tightened the four-point harness as hard as I can during the checklist, and I am glad that I am wearing a parachute. Below and downwind of the lenticular the air rolls around in a turbulent zone called the rotor.

As we climb with the airspeed indicator showing a constant 65 knots the ground speed falls off and I know that, as expected, the wind speed is increasing with height. I can see ragged clouds rotating rapidly and the turbulence thumps both machines up and down and I am using coarse control inputs to stay in formation. We are crawling over the ground now and it is gloomy beneath the lenticular. The sky brightens as we fly out forward of the cloud and suddenly the air is smooth, and the variometer shows 1500 ft per minute climb. I wait 10 seconds and then pull the release and the Cub breaks left and I break right and then back to the left to stay into the wind. At 60 knots indicated with the wings level, I am matching the wind speed, the glider is stationary over the ground and climbing smoothly at 2000 ft per minute. The only noise is the low hiss of the slipstream, and I can fly hands off once I have adjusted the elevator trim.

At 7500 ft I can see dusk falling and break off the climb, open the dive brakes and dive down the front of the lenticular indicating 120 knots. With the brakes open there is a low shrieking noise. When I can see the airfield under the cloud, I close the brakes and glide back with a flat glide angle and high groundspeed because of the strong tailwind. It is the last flight of the day, so I aim to touchdown on the grass close to the hangar. A base leg is flown close to the airfield with a 35° angle to compensate for the wind and cancel out the drift. Turning onto a short final the ground rushes up as I maintain an extra 10 knots because of wind-shear, close the airbrakes to half open and flatten out. The single wheel rumbles and the sailplane is motionless, wings and fuselage level.

In 1972 I made a flight in a Cessna 150 that proves that you don't have to fly a sailplane to soar. In a strong south-westerly wind a line of low mountains creates a standing wave which forms over the town of Middlemarch. It presents as a stationary wall of cloud rising from about 3,000 feet and topping out well above 15,000 feet. I approached it from behind and below and climbed up just forward of the wall. Within seconds the aircraft was climbing at more than 2,000 feet per minute in smooth air and as the lift increased, I throttled back to idle and the rate of climb remained high, four times the Cessna's best rate of climb at full power. Oxygen is needed above 10,000 feet but when I tried to break off at that altitude, I had trouble descending, even with full flap, no power and the maximum airspeed with flaps extended. Eventually, I worked out that the only way out was to retract the flaps, increase the airspeed and fly forward and out of the strongest lift.

Pilots often encounter lift and one day I was with a pupil returning from a cross-country exercise and encountered light wave lift, just enough to maintain altitude with the engine idling. The pupil had an educational experience and I enjoyed something bit different.

CHAPTER SEVEN. ALTERNATIVE AVIATION

In early 1982 I watched three microlight aircraft fly from a farmer's paddock and I was impressed by how simple they are yet could fly under more-or-less perfect control.

The microlight (called ultralights in America) came into being in response to the high cost of flying light aeroplanes. They did this by using the simplest possible structures and the cheapest (but not the most reliable) engines, often from snowmobiles and similar vehicles, which were not designed to run at continuously high-power settings and did not meet the standards of reliability required by aircraft.

The aircraft I watched fly on that day in the early 1980s were startling in their simplicity, better described as minimum aircraft than microlights. They were Australian built Skycraft Scout IIIs, a name more complicated than the aircraft it denoted. The designer was a boat builder with no aviation experience. He researched aircraft and came up with a successful hanglider and in the early 1970s, he built a powered version. It was roughly the size of a small aeroplane, had the same layout but was different in almost every other respect. The main structural member was an aluminum beam, resembling the mast of a yacht and the motor, propeller and fuel tank were attached to the front and the tail at the rear. A plastic lawn chair was attached to an A-frame below the beam and two more beams made up the leading edges of the wings which were made of sailcloth and with a single surface reinforced by battens. The wings were braced by cables and the control surfaces were parts of the wings and tail, which were warped to provide control.

The original Cessna 172 of 1956 weighs 585 kg empty, carry's up to 4 people, has a 145-horsepower motor and cruises at 95 knots. The Skycraft Scout III weighs about 59 kg empty, has a 19 hp motor and cruises at about 35 knots. It carries only the pilot who is out in slipstream and propeller blast. The aircraft was cheap to buy, cheap to run and gets off the ground smartly. However, the small fuel tank and the low cruising speed means it is confined to local flying. It did its job, which was to get people into the air.

The freedom conferred by affordable flying is conveyed in two outstanding books. The first is *Flying for Fun* by Jack Parham, and it describes flights made in 1935-1936. Parham was a middle-aged officer in the British Army's Royal Artillery, who had an 'A' licence but less than 40 hours flying experience when he bought an American built Aeronca C-2, a single seater with a two-cylinder motor of around 26 hp. The C-2 was a little more sophisticated than the Scout, but not by much. Parham flew about 110 hours, learnt a lot, had some fun and enjoyed England from the air.

One day in July of 2012 I flew New Zealand's first and only Aeronca C-100, a two-seat development of the C-2, an aeroplane which had taken my fancy in the 1960s. The Aeronca C-100 is the last iteration of a design which emerged at McCook Field in the 1920s and became a cheap and accessible way to fly at the end of that decade and into the 1930s. The original was a high wing, single-seat monoplane glider using the upper wings from a Curtiss Jenny and was converted to a powered aeroplane using a two-cylinder, air-cooled engine of most unusual provenance, it was designed to pump air into the ballonets of US Army Air Service blimps.

One of these engines was fitted to the glider and the resulting aeroplane was promising. The airframe and engine were developed, and the Aeronca C-2 was the result. The design evolved into the C-3, a popular two-seat trainer in the 1930s. It was then manufactured under licence in the UK as the Aeronca C-100.

In 2012 I went to see an immaculately restored example and thanks to the generosity of the owner and that of a 30,000-hour airline captain, I was able to fly it.

The first thing you would notice is that the high wing aircraft has no struts, instead, it has a king post with bracing to the middle of each wing, and the bottom of the fuselage acts as another kingpost and matching cables go to the underside of the wings. The aircraft looks exactly as if Noddy had designed it.

The first challenge was to get into the cabin. After some experimentation, I found that the only way in was horizontal, with legs and other body parts making their way around the obstacles. It was even more difficult when there was already someone in the other seat.

The two-cylinder air-cooled JAP of around 38 hp provides minimal power and the aircraft (no brakes and no chocks) would not move until full power, a bit of a push and a lift of the tail was applied.

The take-off was an adventure. We used almost all the grass runway and it was not until we were committed that I noticed that the airspeed indicator was not working. The non-sensitive altimeter (the first I had encountered) was a bit hard to read and the airspeed for the best climb was judged by lining up the bottom of the port cylinder on the horizon. After 10 minutes at 50 feet per minute, we were at 500 feet (I think) and settled into cruise at about 55 knots. The 36 feet of wing and large ailerons gave large quantities of adverse yaw, but the glide angle was agreeably flat. The approach and touch-down were easy to judge and the roll-out, with dough-nut tires, gave a smooth ride.

An RAF pilot flew a C 100 from London to Cape Town in the mid-1930s and, although flying the C-100 was a privilege, I don't recommend that anyone emulate him.

The motor of Jack Parham's C-2 self-destructed, and a safe forced landing was made. Rather than fit a new engine Parham converted it to a glider. One of his flights was 1 hour and 5 minutes, soaring in ridge lift. After that, he gave it to a gliding club and was then distracted by preparing for the looming second world war.

Parham comes across as a thoughtful and competent pilot, but the author of *Propellerhead* is neither. Anthony Woodward was an advertising executive and in the 1990s he learned to fly in a microlight owned by a syndicate. He makes no attempt to disguise his lack of aptitude and poor airmanship, but the book is a readable account of the experience of learning to fly and flying, in a very basic flying machine. Readers should be entertained but view the content as a catalogue of how not to fly, as it is what was once called a cautionary tale.

Microlights have resulted in aviation splitting in two. There is the traditional kind, in which most pilots fly factory-built aircraft, train at aero clubs or flying schools and qualify for the pilot licences and ratings which have been the standard for many years. First the Private Pilot Licence, then the Commercial Pilot Licence and Airline Transport Pilot Licence. If they are instructors, they start with a 'C' category then 'B' and then a small number do 'A' category ratings and are regarded as demigods. Alternative aviation has a parallel system of pilot and instructor certification.

In 2010 I gave in to by curiosity and flew a New Zealand built Bantam B22. The cockpit had side-by-side seating for two, a large windscreen but no doors. I was agreeably surprised to find that it had an electric start and wheel brakes. It took off after a short run, climbed out at a reasonable rate and handled in a way that inspired confidence. The instructor demonstrated an important characteristic of an aircraft with low mass and high drag if the engine fails the pilot must lower the nose immediately. If he does not there is a high probability of a stall and spin. I did four or five circuits and enjoyed the experience.

Microlights have mutated, some are very basic, some mimic older types of light aircraft and some are so sophisticated that they are microlights only in the sense that they meet the weight requirements.

I grew up reading about pilots parachuting to the ground from disabled aircraft and I had always wondered what it would be like. The parachutes referred to in the books were the classic round type, made up from gores of white silk but, the one I used was a double surfaced oblong of multicolored nylon. There were eight hours of ground school which concentrated on recognizing a malfunction and what to do about it. This turned out to be grabbing a handle with both hands and pulling it hard and this released the main canopy and opened the reserve.

After waiting months for appropriate weather for a first jump (cloud base above 3,500 feet, surface wind less than 5 knots) I found myself climbing into a Cessna 180 with two other first-timers, jumpmaster and pilot. My ripcord was attached to the aeroplane, so I would not have to remember to pull it. The door had been removed before take-off and at 3,000 feet above airfield the pilot throttled back, I climbed out, using the hand and feet movements I had been taught to use and found myself hanging from the wing strut with both hands. The jumpmaster told me to go and I fell away, looking up and seeing the static line go taut and my weight pulled the parachute out of the bag and I saw the canopy develop.

There was a jerk and a deceleration, and I hung there, moving forward and down, controlling direction with the brake handles. The instructor's voice came over the one-way radio attached to my harness and I pulled left and right until I was lined up for the landing area. The steep descent was easy to judge and pulling half brake and then full brake put me on the ground, standing up and stationary. I looked up and the parachute canopy, was hanging in the air over my head and I pulled down on one of the brake handles until it collapsed in my arms.

I know someone who has done over 14,000 sky-dives but I was happy to note mine in my logbook and leave it at that.

Another thread in the transformation of sports flying is the paraglider and the paramotor. A paraglider looks like a parachute but has developed into a sailplane with the performance to make long soaring flights. The distance record for a paraglider is 564 km, and the altitude record is 27,561 feet, absolutely astounding for a flying machine you can fold up and carry in a backpack. A paramotor is a paraglider with a motor, propeller and fuel tank strapped to the pilot’s back and the records for a paramotor are 24,898 feet and 1105 km. I have seen videos of paramotors doing steep turns with one wingtip on the ground, piling on the power, climbing away, doing a wing-over and repeating the process. Altitude control is good enough to fly with heels in the water, climbing a few feet to go over a low peninsula and descending on to the surface of the water again.

My experience of flying paragliders is only 6 solo and 1 tandem flight but they were fascinating experiences. The wing is laid out on the ground behind the pilot and the lines are checked for damage or tangles. The pilot stands facing into the wind with the lines in each hand. Stepping forward and pulling the lines forward brings the wing up and stable over the pilot’s head. After checking that the wing is fully inflated and the suspension and brake lines are free, the pilot runs forward as fast as possible and becomes airborne. Direction control is by pulling down the brake lines on one side or the other. Speed control is by moving the brake lines up for higher speed and down for slower flight. On landing the paraglider approaches with enough speed to provide the energy for a flare and the brakes are pulled halfway down and then all the way down resulting in a stand-up landing. The experts have remarkably precise control and YouTube features videos of them landing on a balcony railing and doing equally impressive things. My first flights were solo from a hill about 1,200 feet high and I did 2 more from 2,500 feet and I found that you have food for thought when you are dangling below a flexible wing and totally exposed if you misjudge things. The 2 flights I really enjoyed were from a low hill and involved skimming along a few feet above the grass.

I have wondered how much of New Zealand you could cover in a minimum aircraft like the Scout if you had oodles of time, were patient, careful and only flew in perfect weather. I suspect that it would be in stages of about 30 nautical miles each and that weather and maintenance would provide regular delays. If you assume a total distance of 1,000 nautical miles you would have to plan for 35 to 40 stages and there would be times when you would have to pause and return to your life. I would think of it as more of a hobby than a journey. The Scout has one advantage, it can easily be dismantled and stored in any available shed.

CHAPTER EIGHT. TEACHING

An instructor needs to understand what is happening in each phase of each flight and to be able to demonstrate it and explain it in a clear, simple convincing way. The pupil learns something in each lesson and so does the instructor.

One night in February 1977 I was doing circuits at an international airport. With me was a private pilot who was training for his night rating. Descending toward the runway I noticed he was below glide path and not correcting and I asked him to add power, I repeated myself and as we were undershooting the runway and were seconds from hitting the ground, I said 'I have control' took control, landed, taxied in and shut the aircraft down. The pupil was clearly pouting. He told me I had been rude. I pointed out three things, I was pilot in command, I could not let the aircraft crash and if there was an interpersonal issue it should be discussed post-flight. He was a C-list celebrity and let his pride override his instinct for self-preservation.

I was 22-years-of-age when I started instructing and it was an eye-opening exposure to just how challenging some people could be. Sometimes there was an ordinary clash of personalities, nobody's fault, sometimes there were strange interactions. There was a university lecturer who was fond of pompously saying that the instructor was his guest rather than his instructor. I remember a farmer, a very nice chap who had never worn a watch and I had to teach him how to use it in navigation. Some pupils had no aptitude and insisted on taking lesson after lesson, despite clear evidence that they would never be skilful enough to fly solo. Some pupils had good stick-and-rudder skills, but their judgement was poor, and they were resistant to advise.

Flight time to the first solo is an index of aptitude. Pupils who fly at least once per week and experience consistent fine weather solo in between 8 and 12 flying hours.

The longest I am aware of was 30 hours and another took 24 hours. Coincidentally these two learners were doctors in the same medical practice. When I started to instruct, I found that aeroplanes were a great leveller, achievement in other areas of expertise was no guarantee of achievement in aviation. I remember an 18-year-old apprentice plumber (a nice chap but not super bright) going solo in seven hours when university lecturers sometimes took twice that long. A friend of mine took 3 hours and 55 minutes which is outstandingly impressive. In the early years of flying, before and during the First World War pupils often went solo with only a few hours, which is one of the reasons that there was about one fatality in training for every death in combat.

I learned enough about aerobatics to teach them but did not do them with any enthusiasm. Most of my aerobatic experience was in a Beagle Pup, an aircraft with light well-harmonized controls and a reasonably powerful engine. Modern aeroplanes are designed to have safe stalling characteristics and the Pup was no exception. Lead weights could be attached to the ventral fin for training in spin recovery. With the weights, it entered a spin cleanly and would keep on spinning until the correct recovery procedure was used.

Flying is flying but instructing is an art as much as a science. Understanding what you are doing is not the same thing as being able to explain it simply and clearly. The instructor must understand and communicate essential concepts such as the difference between the aircraft's path through the air being different from its path over the ground when there is wind aloft. One day I was on a dual cross-country when we flew into an air mass moving at 45 knots from left to right across our track. To keep flying along our chosen track, I had to adopt a heading well to the left of it. The aircraft was still flying cleanly through the air but the allowance for drift meant that we had to view the track through the right-hand side of the windscreen and the window in the right-hand door. I was able to explain the phenomena clearly and simply to the student pilot at the post-flight debriefing. My source was Stick and Rudder: An Explanation of the Art of Flying by Wolfgang Langeweische noted in the next chapter.

There is no rocket science in the operation of light aeroplanes but there is a lot of information to acquire and apply. An instructor must be able to sketch out simple diagrams to aid in the explanation of technical points and flight manoeuvres. When I started flying and studying its technical aspects, I found that the context made the study interesting in a way that was absent in general science at school.

CHAPTER NINE. THE VIEW FROM ABOVE

One evening I met a woman at a party and offered to take her flying. Before dawn on the next morning, there was a knock on the door and there she was. We drove out to the airport and I opened the hangar doors, pulled out a Piper Cherokee and pre-flighted it. I started up right at the start of morning civil twilight and got a taxi clearance. As I lined up on the runway it was getting light, took-off and climbed out to the west, over green valleys filled with early morning fog. As we climbed the sun appeared to us although it had not yet become visible to those on the dark ground below. With full daylight, we climbed to 10,000 feet in air so still that the Cherokee would fly hands off for minutes at a time. It was one of those rare autumn days when the air was perfectly clear, and we could see many miles in every direction.

One of the delights of flying yourself is that you see things that groundlings don't. The meanders of an old river, with its loops and ox-bow lakes, are just markings on a map but beautiful from the air. Changes in altitude and lighting mean that you often see the same scene in different ways. The ground seen through gaps in the cloud deck gives the pilot a sense of detachment from earthbound things.

When I was at law school in the late 1970s, I decided that I would see as much of the country as possible and got a job flying from a different base each summer. The pay was bare subsistence, but my days were rich in natural beauty.

I explored every river, lake, road, airfield, mountain, beach and hill within reach. One day I flew into and out of the crater of a steaming volcano. I experienced my native land as only a pilot can. One day I was flying with the local police officer on a search for a missing boat. As we covered every nook and cranny of a complex coastline, we saw a hammer-head shark, as long as a small boat, swimming in clear water, only a few hundred meters from a bay lined with half-a-dozen holiday homes.

These holiday jobs usually lasted about two months each, but in January 1980 one lasted exactly a day. An operator had three four-seat Maules and carried tourists over some of the roughest and most beautiful landscapes in the county. He had employed a youngish and relatively inexperienced pilot and needed a third man over the busy holiday period.

I drove up from my home and arrived to find that the youngish guy had pranged one of the Maules, and there were only two aircraft to fly. The first thing the next morning the boss and I set out to convert me to the Maule and familiarise me with operations. I did two complete trips with him in the right-hand seat and three tourists sharing a back seat intended for two people.

The Maule turned out to be a nice aircraft to fly, it handled well and had good short take-off and short landing characteristics. Each flight was a climb up to a mountain covered in ice and snow and almost 10,000 feet high, over glaciers and down into an airfield by a fjord. After a photo opportunity and swatting a few bugs we took-off again and flew over steep rocky slopes and to an airstrip in a deep valley. It was still early in the morning and the valley was covered in thick cloud. The boss took over and flew back and forward until he saw a hole in the cloud, a very small hole, rolled into a maximum rate turn while descending steeply. There were shrieks from the girls in the back and he turned, smiling and told them there was no extra charge for the thrills. We popped out below the cloud base and I took over.

The strip was in a tight bend in the river and had a section of the river at each end. There was no windsock and a pilot hand to gauge the wind direction from wind-lanes on the water. It was essential to get the wind direction right as a down-wind landing would put the aircraft into the river. I landed, the girls got out and walked to a spot on the riverbank where they were picked up by a jet boat.

We had done the first flight in loose formation with another aircraft and I was shown how to stay in formation by using the controls and not using the throttle, which would have increased the fuel consumption. I noticed that we did not refuel after the flight because the fuel consumption was higher with the tanks full, although that meant the safety margin on the next flight would be slim.

I did the second flight over the same route, the cloud over the river airstrip had dispersed but the wind on the ground was from the same direction. It was decided that was checked out on the aeroplane and the airfield and airstrip and was told to take one of the Maules down to another airstrip and do an hour's take-offs and landings solo. I would take out my first load of passengers at dawn the next day. But it did not work out like that. The boss headed off to do a job in a helicopter and I waited for my Maule to arrive.

By the time the boss arrived back, the Maule was two hours overdue. We pulled the third Maule out of the hangar and flew up the valley towards the airstrip in the bend. The missing aircraft was sitting at one end of the airstrip, at right angles to the runway. The wind had changed, and the pilot had not noticed, landed downwind and had to ground loop to avoid going into the river. There was no-one around and we inspected the Maule. The port wing had dug in, smashing the tip and wrinkled the skin on the top of the wing, the port elevator was bent up at a 30-degree angle and the tire on the port undercarriage leg was stripped off the hub. It was unflyable and would need major, time-consuming repairs.

I assumed that the aircraft would be dismantled and helicoptered out in pieces, but I was wrong. Towards nightfall, I was at the base airfield when I heard an aircraft approaching. It was the damaged Maule. The boss had smashed the fiberglass of the port wingtip to free-up the aileron on that side, sawed the damaged portion of the elevator off, replaced the tire and flown it out.

It had been a fascinating experience and I was glad I had had it, but it was a hazardous operation and I was well out of it.

With no aircraft for me to fly I was out of a job and spent the rest of the summer instructing at a flying school.

CHAPTER TEN. BOOKS

The literature of aviation is vast but much of it is highly specialized and many books will only interest the reader interested in the minutiae of the history of a squadron or an aircraft manufacturer or a type of aeroplane. The books described below are for the flyer and non-flyer alike.

Not all flyers are readers but if they aren't, they are missing out on something entertaining and educational. Here are details of the books I have mentioned earlier in this book and other first-class aviation books, gleaned from a lifetime of reading on aviation topics. They are essential reading for anyone who wants to immerse themselves in the aviation experience.

The cornerstone work on how and why an aeroplane flies is A. C. Kermode's *Mechanics of Flight* (12th Edition, 2012). The same author wrote *Flight Without Formulae* (1941), a classic which makes the principles of flight accessible to the beginner. The FAA *Pilot's Handbook of Aeronautical Knowledge* (2016) contains answers to all the questions a new pilot might ask.

Wolfgang Langewiesche's *Stick and Rudder: An Explanation of the Art of Flying* (1944) is the best summary of the information all pilots need-to-know and is still in print (McGraw Hill). Wolfgang Langeweische had a son called William who was a career pilot from his teens to his thirties and is now a successful writer of non-fiction including some pieces about aviation. Father and son can convey complex information in a simple and entertaining way. *Aloft: Thoughts on the Experience of Flight* (2010) includes 'The view from above' about the unique perspective on the earth obtainable from an aeroplane and 'The angry sky', about deliberately flying in bad weather. His *Fly by Wire: The Geese, the glide and the miracle on the Hudson* (2010) is an insightful account of airline flying in the digital age and the incident which left alive all 155 persons on board.

Ernest C. Gann was an airline pilot from the late 1930s to the 1950s and his 10,000 plus hours of flight time included many memorable incidents. His *Fate is the Hunter*

(1961) is a unique description of a flying career rich in incident. The book is a must-read for anyone interested in the days of the Douglas DC-2, DC-3 or DC-4. In between extreme weather events, Gann had to deal with multiple engine failures. On one occasion three of four engines on his DC-4 lost power just after lift-off and the circuit was made at 50 feet. The flight, approach and landing were made with full power. Selecting gear down and flaps down caused a descent and he touched down without reducing power. While Gann is famous for *Fate is the Hunter*, he was a successful novelist whose work includes the aviation stories; *Island in the Sky* (1944) *Blaze at Noon* (1946), and the World War One Novel *In the Company of Eagles* (1966).

Charles Lindbergh was the consummate professional pilot and his book *The Spirit of St Louis* (1953) tells two stories in parallel. The first is how he learnt to fly and became a skilled aviator and the other is his solo flight from New York to Paris in 1927. The learning part includes teaching himself most of what was to be learned about flying, his time as a parachutist, barnstorming in Curtiss Jennys, honing his skills with the United States Air Service and flying the mail, winter and summer in all weathers. He spent 15 years writing and revising it and it won a Pulitzer Prize in 1954.

Jack Parham's *Flying for Fun: An Affair with an Aeroplane* (1983) is my favorite book about the sheer pleasure of being aloft with no intention except to enjoy oneself. BBC 2 made a movie based on Parham's book in the 1980s and it is available on YouTube. Almost as good as *Flying for Fun* are Harald Penrose's *No Echo in the Sky* (1958) and *Airymouse* (1968). Penrose was something of a renaissance man, an engineer, test pilot, naval architect, historian and author. His autobiography, Adventure with Fate (1984) covers his flying life from the 1920s to the 1970s and includes testing many prototype aeroplanes, as the chief test pilot for Westland Aircraft. *Propellerhead* by Anthony Woodward (2002) is well written and entertaining but an example of how not to fly microlights.

Terence Horsley's book *Soaring Flight* (1944) is a snapshot of the state of gliding art at the outbreak of World War Two in 1939, while George Moffat's *Winning on the Wind* (1974) is a readable book by a US and World champion glider pilot, covering his gliding from beginner to champion. Helmut Reichmann's *Soaring Cross Country* (7th Edition, 1993) is a classic by another world champion. Philip Wills was world champion in 1953 and his *On Being a Bird* (1953) and *Where No Birds Fly* (1961) are readable classics. Carle Conway's *The Joy of Soaring: A Training Manual* (1970) conveys the basics of glider flying with text and clear diagrams.

Richard Bach is a pilot who has many best-selling books and my favorites are his earlier works including; *Stranger to the Ground* (1963), *Biplane* (1966) and *Nothing by Chance* (1969) which describes flying Republic F84F fighters in the United States Air Force, being a modern barnstormer and generally experiencing the magic of flight.

Jeffrey Quill tested every model of Supermarine Spitfire and his *Spitfire: A Test Pilot's Story* (1983) is readable and unique. My favorite chapter is about flying open cockpit biplane Siskin fighters in the meteorological flight in the mid-1930s. This entailed daily climbs to above 20,000 feet, taking readings of pressure and temperature and cloud types. He had heated clothing, oxygen, and the bare minimum of blind flying instruments.

Anthony Smith's book about old style ballooning is a classic. Smith was a biologist, author, and traveller. In 1960 he decided to celebrate the 100th anniversary of the publication of Jules Verne's *Three Weeks in a Balloon* by learning to fly a classic hydrogen balloon and taking his own balloon to East Africa. The story works as an adventure and is a very good introduction to the fine old art of flying silently beneath 26,000 ft3 of hydrogen rather than the noise and heat of the burners of a modern hot air balloon. He tells the story in *Jambo: African Balloon Safari* (1963) and *The Dangerous Sort: The Story of a Balloon* (1970).

Everything you might want to know about flying airships in the First World and up to 1930 is included in the memoirs; *My Airship Flights: 1915-1930* (1970) by George Meager and *Airship Pilot Number 28* by T. B. Williams (1974). Nevil Shute's *Slide Rule: The Autobiography of an Engineer* describes the design, building and flying of R100, one of the last large rigid airships.

James Salter's *The Hunters* (1956) is an autobiographical novel about fighter pilots in the Korean War of the early 1950s is the most convincing portrayal of the aviator as a warrior I have ever read.

Many classic aviation books will be available in your local library and any of the standard bookselling sites on the internet. These two aviation bookshops have an outstanding range of aviation books:

Aviation Megastore (Netherlands):
https://www.aviationmegastore.com

The Aviation Bookshop (Great Britain):
https://www.aviation-bookshop.com/

The reader may access many classic aviation books free though: HathiTrust, Internet Archive and Project Gutenberg and similar sites. If you search the internet for a book and give the author and title you will usually find multiple sources for the book.

APPENDIX ONE
1956 & 2010 CESSNA 172

The 172 is the benchmark by which other light aeroplanes are judged and here are the brochure figures for the 1956 and 2010 models. Like all brochure figures, they are optimistic and would not be used in practice.

A safety conscious pilot would probably assume a cruising speed of about 110 mph and a reduced range when flight planning and would intend to land with fuel for at least 45 minutes.

The 1956 model of the 172 had a six-cylinder Rolls-Royce Continental engine and it ran very smoothly and quietly. It was replaced in later models with a four-cylinder engine that was just as reliable and slightly more powerful but ran noticeably rougher.

All Cessna 172s have four seats and are fitted with dual controls. They can be fitted out for a wide range of roles.

The factory figures for the 1956 Cessna 172.

Stalling speed flaps extended 52 mph.
Stalling speed flaps retracted and power off 58 mph.
Maximum speed. 131 mph.
Cruising speed. 125 mph.
Vne (never exceed speed). 160 mph.
Range 491 miles.
Endurance 3.9 hours.
Rate of climb at sea level 660 feet per minute.
Service ceiling 15,100 feet.
Take-off distance (to 50 feet altitude) 1650 feet.
Landing distance (from 50 feet altitude) 1115 feet.

Gross weight 2,200 pounds.
Empty weight 1290 pounds.
Useful load 910 pounds.
Wing loading 12.6 pounds per square feet.
Power loading 15.1 pounds per horsepower.
Fuel capacity 39 US gallons.
Engine. Rolls-Royce Continental 0-300 of 145 horsepower.
Propeller. Fixed pitch metal 76-inch diameter.

The factory figures for the 2011 Cessna 172R

Length: 27 ft 2 in (8.28 m)
Wingspan: 36 ft 1 in (11.00 m)
Height: 8 ft 11 in (2.72 m)
Wing area: 174 sq ft
Aspect ratio: 7.32
Airfoil: modified NACA 2412
Empty weight: 1,691 lb (767 kg)
Gross weight: 2,450 lb (1,111 kg)
Fuel capacity: 56 US gallons (212 litres)
Powerplant: 1 × Lycoming IO-360-L2A 4-cylinder, 160 hp (120 kW)
Propellers: 2-bladed metal, fixed pitch
Cruise speed: 122 knots (226 km/h)
Stall speed: 47 knots (87 km/h) (power off, flaps down)
Never exceed speed: 163 knots (302 km/h)
Range: 696 nautical miles (1,289 km) with 45-minute reserve, at 55% power.
Service ceiling: 13,500 ft
Rate of climb: 721 ft/min
Wing loading: 14.1 lb ft2.

The full specifications provided by Cessna for their 2010 Cessna 172R are at:

file:///C:/Users/garth/Pictures/how%20to%20fly%20a%20plane/2010%20Cessna %20172R.pdf

APPENDIX TWO. GLOSSARY

This glossary is the product of my years of flying and teaching others to fly and is based on the language used in the country in which I fly, which is New Zealand. The book uses British English, which differs in some ways from English as written elsewhere. If you are flying in the United States or Canada you will find that some of the terms are spelt a little differently, but you will have no problem adapting.

Adverse Yaw. The nose of the aircraft yaws away from the direction of roll in a turn.

Aerobatics. There are four aerobatic maneuvers: roll, loop, stall turn and spin.

Aerofoil. A surface which produces lift, mainly the wing.

Aeroplane. Powered heavier than air aircraft.

Aerotow. Launching a glider by towing behind a light aeroplane.

AGL. Above ground level.

Aileron. Control surfaces hinged to the rear of each wingtip and providing control in roll, around the aircraft's longitudinal axis.

Aircraft. All aircraft are either aerodynes (heavier than air) such as aeroplanes and gliders or aerostats (lighter than air) such as balloons and airships. In this book, I have used the word to denote an aeroplane or glider.

Airframe. The structure of the aircraft. Usually used to describe the aircraft without the engine or propeller.

Airship. A steerable balloon.

Airspeed. Speed through the air. Not to be confused with groundspeed.

Airspeed indicator. A pressure instrument which shows the airspeed often abbreviated as **ASI**.

Altimeter. A pressure instrument showing altitude. It can be set to show either altitude above ground or altitude above sea level.

Attitude. The inclination of the axes of the aircraft to the horizontal. For example, an aircraft may be in a nose up (above the horizon) attitude.

Angle of Attack. The angle between the chord line of the wing and the relative airflow.

Anhedral. A negative angle between the horizontal and the wing, when viewed from the front.

AGL. Above ground level.

ASL. Above sea level.

Aspect Ratio. The ratio of span to chord. Gliders are high aspect ratio because this reduces induced drag and improves glide ratio.

Axis. A line through the center of gravity of the aircraft in each major plane. Longitudinal is fore and aft. Lateral is crosswise. Vertical is up and down.

Ball. A ball in a curved tube and which shows whether an aircraft is slipping, skidding or in balanced flight.

Balloon. An un-powered lighter-than-air aircraft inflated with either hot air or helium or hydrogen.

Barograph. A device carried in a glider to record altitude and time. The resulting trace is evidence used when setting records or qualifying for gliding awards

Best Glide Speed. The indicated airspeed at which a glider or powered aeroplane with its engine stopped achieves its flattest glide.

Buffeting. A disturbance in the airfoil around the aircraft indicating that the wing is close to its stalling angle.

Bungy Launch. A type of launch in which a glider is accelerated to flying speed by a V-shaped elastic rope.

Cabin. In a small aeroplane, the cabin houses the pilot and the pupil or pilot and passengers.

Camber. The curve of the upper surface of the wing.

Canopy. (1) The clear plexiglass covering of a cockpit. (2) The part of a parachute which provides the drag and slows the descent.

Cantilever Wing. A wing with no external bracing wires or struts.

Centre of Gravity. The point on aircraft that its weight is assumed to act. For a safe flight, the center of gravity must be within a prescribed range.

Centre Section. The middle portion of a wing.

CFI. Chief flying instructor.

Chord. A straight line from the leading edge to the trailing edge of a wing.

Circuit. The airspace above an airfield. Used for planning and carrying out take-offs and landings.

Clean. An aircraft is said to be clean when the undercarriage, flaps, slats and any other retractable device, are retracted.

Cockpit. Part of an aircraft which houses the pilot and pupil or pilot and passengers.

Control Surfaces. Hinged surfaces which control the aircraft around the three axes. Rudder for yaw, ailerons for roll and elevators for pitch.

Convection. The upwards movement of part of the atmosphere. Usually referring to thermal action.

Coordination. The use of the control surfaces to achieve the desired flight path.

Course. The heading which causes the aircraft to fly along the desired track.

Control Wheel. The cockpit control which operates the ailerons and elevators. Also called the stick (although it is rectangular), and less often, the yoke.

Crossed Controls. Simultaneous use of ailerons and rudder in opposite directions, causing a slip or skid.

Dihedral. The positive angle between each wing and the horizontal when viewed in front elevation. The reverse is called anhedral.

Dive Brakes. Flat plates folded into the wings of sailplanes which are extended to increase drag and steepen the glide angle, usually when approaching to land.

Drag. One of the forces created by the passage of an aeroplane through the air. The force opposing the passage.

Induced drag is created by the wing as it produces lift.

Parasitic drag is produced by parts of the aircraft that do not produce lift.

Dual Control. Two sets of controls, one for the pupil and one for the instructor.

Dual Instruction. A trainee pilot flying with an instructor.

DVA. Drill of vital actions. A list of items checked before taking off.

Dynamic Lift. The lift produced by the passage of an aerofoil through the air. An airship flown at an angle to the airflow generates dynamic lift in addition to its static lift.

Effect of Controls. The pupil pilot's first flying lesson in which the instructor demonstrates the controls and has the pupil get the feel of the stick and rudder, the effect of opening and closing the throttle and the other characteristics of the aeroplane.

Elevator. A control surface hinged to the rear of the horizontal tailplane and producing nose up and nose down movement.

Envelope. A fabric bag consisting of layers of doped fabric, containing the lifting gas in a balloon or airship.

Fin. The fixed, vertical part of the tail.

Flaps. Surfaces hinged to the rear edges of the inboard portion of the wing. They provide extra lift and drag.

Flare Out. Flattening the glide and reducing speed for landing

Flight Manual. A book carried in all aircraft setting out essential information. This includes limiting speeds, the centre of gravity range, empty and all-up weights and how to calculate take-off and landing distances.

Floatplane. A seaplane with floats.

Flying boat. A seaplane with a fuselage which doubles as a boat. Most flying boats also have tip floats, to prevent the capsizing sideways.

Follow me through. The instructors command when he or she flies the aeroplane with the pupil's hands and feet on the control wheel and rudder pedals.

Fuselage. The body of an aeroplane or glider.

G-Forces. The forces experienced by an aeroplane in flight, measured in multiples of the force of gravity. It is one G during straight and level flight and more in a turn or recovery from a dive or during aerobatic maneuvers. The force may be positive or negative. A standard light aeroplane is stressed to withstand about 3.5 G positive and 1.5 G negative. A purpose-built aerobatic aeroplane is stressed to withstand much greater G forces.

Glide Ratio. The ratio of distance covered to altitude lost. An aircraft which is covering 20 feet forward for every 1 foot of altitude lost has a glide ratio of 20 to 1.

Glider. An aeroplane without a motor.

Ground Loop. Loss of directional control while landing, taking-off or taxiing.

Groundspeed. The speed relative to the ground, but not relative to the air. With a tailwind the groundspeed is the airspeed plus the wind speed, the reverse is true with a headwind.

Form Drag. The drag created by the shape of the fuselage, wing and other parts of an aeroplane or glider.

Heading. The direction that the aircraft is pointed.

High Tow. A sailplane flying above the wake of the tow plane is said to be in high tow.

Hot-air-balloon. A balloon inflated with hot air. **Gas balloons** are inflated with hydrogen or helium.

I have control. The instructors command when taking control from the pupil.

Induced Drag. The part of the drag which is incidental to the creation of lift.

Knot. One nautical mile per hour.

Laminar Flow. Smooth airflow across an aerofoil

Landing Configuration. If an aeroplane is fitted with retractable undercarriage and flaps, it is in its landing configuration if both are extended. If only flaps are fitted, they are extended during approach and landing.

Landing Speed. The indicated airspeed at which an aircraft touches down when landing.

Leading Edge. The front of the wing.

Lenticular Cloud. A cloud form shaped like a lens or aerofoil, indicating the presence of wave lift.

Lift. (1) The vertical component of the forces created by the passage of the wing through the air. (2) Rising air which allows a sailplane to maintain height or gain height.

Lift Drag Ratio. The ratio of lift to drag, abbreviated as LD or L/D.

Lift Strut. A strut bracing the wing to the fuselage. Used in aeroplanes such as the Cessna 172.

Load. The forces acting on an aircraft. The **static load** is the force of gravity when the aircraft is on the ground. **Manoeuvring loads** are produced using the controls. **Gust loads** are produced by turbulence in the atmosphere. Total loading is the sum of all three.

Load Factor. The sum of the forces acting on the aircraft. All aircraft have limiting load factors which are expressed as multiples of the force of gravity i.e. G-forces.

Loop. An aerobatic maneuver in which the aircraft turns through 360 degrees in the vertical plane.

Low Tow. A sailplane flying below the wake of the tow plane is said to be in low tow.

Manoeuvring Speed. The speed below which the controls can be used fully without risking damage to the airframe. Usually stated on a placard in the cockpit with other essential data.

MAUW. Maximum all up weight. The total weight of the airframe, fuel, pilot, passengers, and cargo, which must not be exceeded.

Microlight. A very light aeroplane which is often of simple construction with a low all-up-weight and stalling speed.

Minimum Sinking Speed. The indicated airspeed at which an aircraft descends at its lowest rate.

Motor-Glider. A sailplane with a motor for launching the sailplane. When the pilot finds lift, he or she will stop the motor and feather the propeller. If the lift ceases the motor will be re-started for the return to the airfield.

Nacelle. The casing of an engine or another component. Some microlights have fuselages which are a pod (i.e. nacelle) for the pilot and passenger and a boom with the engine at the front and the tail at the back.

Navigation Lights. When flying at night aeroplanes show a red light on the left wingtip, a green light at the right wingtip and a white light at the tail

Nautical mile. One 60th of one degree of latitude. 6080 feet. Used in flight navigation.

Overshoot. Either touching down further down the landing area than intended or applying power to climb and circle the landing area and try another landing.

Paraglider. A glider with a rectangular foldable fabric wing. It consists of a double surface wing, suspension lines and control lines and a harness for the pilot which, has a simple seat built into it. It is foot-launched. The whole paraglider folds into a backpack.

Paramotor. A paraglider plus a power unit strapped to the pilot's back. The power unit consists of a motor, propeller, fuel tank and throttle. It is usually foot-launched.

Parasite Drag. All drag which does not create lift.

Paratrike. A paramotor with a three-wheeled undercarriage and a seat.

Penetration. The distance a sailplane makes good into a wind.

Pitch. The motion of an aircraft; nose up or down around its lateral axis and controlled by the elevators.

Pitch Attitude. The place of the nose, on, above or below the horizon.

Pitot Head. The Pitot Head faces into the airflow and receives the dynamic pressure of the airflow. The airspeed indicator works by comparing the static plus the dynamic pressure of the air with the static pressure.

Port. Left.

Powered Parachute. A large fabric wing and rigging, resembling a parachute or paraglider, with a rigid unit consisting of two side-by-side seats, engine, propeller, fuel tank and engine controls. A paramotor by another name.

QNH. Altimeter setting to show altitude above sea level.

QFE. Altimeter setting to show altitude above the airfield.

Towline Release. A red toggle in the cockpit of a sailplane that releases the tow rope.

Retractable Undercarriage. An undercarriage which is folded into the aircraft when it is in flight.

Ridge Lift. Rising air found on the windward side of a ridge in which a glider may soar.

Roll. (1) The motion of an aircraft around its longitudinal axis. Aeroplanes roll when the ailerons are used. (2) In aerobatics a turn through 360 degrees in the horizontal plane.

Roll Out. The part of the landing sequence from touching down to rolling to a stop.

Rotor. An area of turbulent air under a wave cloud.

Rudder. Control surface hinged to the vertical fin and providing control in yaw (the nose left or right).

Rudder Bar. A foot operated control which performs the same function as rudder pedals. Usually found in old aeroplanes

Rudder Pedals. Foot operated control that moves the rudder to yaw the nose left or right. In Cessna 172s and similar aircraft, the rudder pedals steer the nosewheel for taxing and pushing the tops of the pedals causes differential braking.

Sailplane. An aeroplane without an engine, capable of soaring.

Seaplane. An aeroplane designed to take off and land from water.

Sea Breeze Front. An area of rising air produced by the meeting of the warm inland air and the moist cool air from the ocean.

Side-by-side Seating. The pilot and instructor (or front seat passenger) sit side by side, as in a car. The convention is that the pilot or pupil sits in the left-hand seat and the instructor or passenger sits on the right.

Sink. An area of air which is descending.

Slide-slip. Increasing the rate of descent by crossing the aileron and rudder controls.

Slot. A gap between two parts of a wing which smooths the airflow and increases lift.

Soaring. Maintaining height or climbing without power.

Solo. Flying alone, usually used to refer to a trainee pilot flying with a flying instructor.

Span. The distance between the left and right wingtips of an aircraft.

Spin. An aerobatic maneuver in which an aircraft descends steeply at a low forward speed, with the wings stalled and auto-rotating.

Spiral. A descent combined with a turn.

Spoilers. Simple dive brakes fitted to the top surface of the wing of a glider. They are flush with the wing when not in use.

Sponsons. Wing-like stabilisers mounted on the hulls of some flying boats and used to keep them upright when afloat.

Stability. The tendency of an aircraft to return to its attitude after being disturbed by control movement or air movement.

Stall. When the airflow over the wing breaks up the wing is said to be stalled.

Stalling speed. The airspeed at which the wing stalls in steady straight flight. The stalling speed is higher when the load is more than 1 G, such as in a steep turn or a dive recovery.

Stall turn. An aerobatic manoeuvre made up of a vertical climb followed by a turn through 180 degrees in the vertical plane followed by a vertical descent.

Standard atmosphere. A hypothetical set of figures including temperature, pressure, density and the rate at which temperature, pressure and density reduce with increasing altitude. Used for calculating performance figures.

Static Vent. Measures the static pressure of the air for the altimeter and the airspeed indicator.

Static lift. The lift produced by the gases in the envelope of a balloon or airship.

Static Line. The line attached to a parachute and the jump plane, which automatically opens the parachute.

Starboard. Right

Step. A v-shaped feature on the underside of a seaplane hull or float.

Stick. A lever which operates the elevators and ailerons. Often used to describe a wheel or yoke which perform the same functions. Usually in front of the pilot but sometimes a stick is mounted between the pilots.

Tail. The rear section of an aeroplane i.e. tailplane, elevator, fin and rudder.

Tail-down Undercarriage. A type of undercarriage which features two wheels forward and a smaller wheel or skid under the tail. The aircraft is tail-down when on the ground.

Tailplane. Fixed horizontal section of the tail surfaces.

Tandem Seating. The pilot and passenger seats are one-behind-the-other. Typical of older aeroplanes. In vintage aeroplanes with tandem seating, the instructor sat in the front seat and the pupil in the back.

Thermal. An area of rising air caused by differential heating of part of the terrain and in which a sailplane may circle and soar.

Touch and Go. A landing followed and take-off without stopping.

Three Axes. An aircraft moves around three axes: vertical, horizontal and longitudinal. The aircraft yaws left and right about the vertical axis, pitches nose up and down about the horizontal axis and rolls left and right around the longitudinal axis.

Three Axis Control. An aircraft which has ailerons, rudder and elevators is said to have three-axis control.

Track. The path of the aircraft over the ground.

Tricycle Undercarriage. The third wheel is under the nose and the aircraft is level when on the ground.

Trim. To cancel unwanted control forces by using trimming devices e.g. trim tab or adjustable control surface or a spring in the control circuit. Simple aircraft have elevator trim while more complicated aircraft may have elevator, rudder and aileron trim.

Trim Tab. A small aerofoil attached to a control surface.

Trim Wheel or Trim Lever. The control in the cockpit which operates the trim tabs or other devices used for trimming the aircraft.

True Airspeed. Indicated airspeed corrected for temperature and pressure.

Turn and Slip Indicator. An instrument which shows the rate of turn and whether the aircraft is slipping or skidding. A needle shows the rate of turn and a ball in a tube shows slip or skid.

Two-Axis Control. An aircraft which has a rudder and an elevator, but no ailerons is said to have two axis controls.

Variation. The difference between geographic north and magnetic north. This is noted on aviation charts and it varies from place to place and over time.

Variometer. An instrument which shows the rate at which a sailplane is rising or descending. Synonymous with the climb and descent indicator and vertical speed indicator.

Vertical Speed Indicator. An instrument which shows the rate at which an aeroplane is climbing or descending. A synonym for **variometer.**

Wake Turbulence. An area of disturbed air caused by the passage of an aircraft through the air.

Wave Lift. Rising air caused by wind flowing over a line of mountains and producing one or more areas of rising air downwind.

Weak Link. A section of a sailplane towrope designed to break between a specified minimum and maximum load

Weight Shift Microlight. A hang glider wing attached to a tricycle undercarriage unit to which seat(s), engine and fuel tank are attached, and which is controlled by the pilot shifting his weight.

Wheeler. A landing in a tailwheel or tail skid equipped aircraft in which the aircraft touches down on the main wheels, in a level attitude and holds that position until the tail drops later in the landing roll.

Winch Launch. A glider is linked to a cable which is attached to a winch at the far end of the airfield. The winch reels in the cable and the glider is pulled along the ground and into the air.

Wing. A fixed surface of aerofoil section which provides an aeroplane with its lift and a mounting for the ailerons. A monoplane (single wing) is said to be made up of the port (left) and starboard (right) wings.

Wing Loading. The weight of an aircraft divided by the wing area.

Wing Span. The distance between the left and right wingtips.

Wing Tip. The outer part of a wing.

Yaw. The motion of an aircraft around its vertical axis, nose left or nose right, controlled by the rudder.

Yaw String. A thread taped to the canopy of a sailplane, which shows the pilot whether the sailplane is slipping, skidding or in balanced flight.

You have control. The instructors command for the pupil to take-over the controls.

Made in the USA
Monee, IL
04 November 2026

67352879R00030